STUDENT GUIDE

FROM THE GROUNDUP

MODELING, MEASURING, AND CONSTRUCTING HOMES

MathScape™
SEEING AND THINKING MATHEMATICALLY

To: Apprentice Architects
From: From the Ground Up, Inc.

Welcome to From the Ground Up. As an apprentice architect, you will be involved in all the stages of designing model homes: drawing floor plans, constructing scale-size walls and roofs, and estimating costs. To help you design the homes, you will be learning about geometry, measurement, scale and proportion, and cost calculations.

What math is involved in designing and building a model home?

FROM THE GROUND UP

PHASE**ONE**

Floor Plans, Site Plans, and Walls

In this phase you will be using scale and metric measurement to design a floor plan, place the floor plan on a building site, and make outside walls with doors and windows. Next you will determine the area of the rectangular walls. Finally, you will calculate the cost of building the walls, based on area.

PHASE**TWO**

Roofs, Area, and Cost

You will experiment with ruler, scissors, compass, and tape to draw "nets" for four roof model types. After gaining expertise in roof design, you will design a roof for the house model you began in Phase One. You will end the phase by determining the area and cost of materials for your roof, floor, and ceiling.

PHASE**THREE**

Budgeting and Building

You will complete a cost estimate for your home design combining the estimated labor costs with the estimated costs for materials. By looking at costs for your classmates' model homes, you will see how cost and design are related. You will apply what you have learned in the unit to a final project: designing a home within a budget.

PHASE ONE

To: Apprentice Architects

Your first assignment is to plan and build a model home for the new community of Hill Valley. The Hill Valley developers want all the homes to be one story, but they also want a variety of floor plans so the houses don't all look alike. They don't want any rectangular floor plans.

We would like you to begin by designing a floor plan that meets Hill Valley's Building Regulations and Design Guidelines. We will soon be giving you more information about the walls, ceilings, roof, and costs. We're eager to hear your ideas for this project.

In this phase you will be designing a floor plan for a model home. By using scale, you will be able to represent the sizes of rooms and furniture on paper.

Making and understanding scale drawings and models is an important skill for many professions. Mapmakers, architects, and set designers all use scale. What other professions can you think of that use scale?

Floor Plans, Site Plans, and Walls

WHAT'S THE MATH?

Investigations in this section focus on:

SCALE and PROPORTION

- Figuring out the scale sizes of actual-size objects
- Figuring out the actual sizes of objects based on scale drawings
- Creating scale drawings

GEOMETRY and MEASUREMENT

- Estimating and measuring lengths
- Estimating and measuring areas of rectangles
- Comparing different methods for finding the area of rectangles

ESTIMATION and COMPUTATION

- Making cost estimates

MathScape Online
mathscape2.com/self_check_quiz

Designing a Floor Plan

CREATING AND INTERPRETING SCALE DRAWINGS

The first step in designing a model home is to create a floor plan. You will need to think about how to design a floor plan that meets both the Guidelines for Floor Plans and the real needs of people who might live in the home.

What would your classroom look like in a scale where one centimeter represents one meter (1 cm:1 m)?

Relate Scale Size to Actual Size

Comparing the Guidelines for Floor Plans to the room you are in and objects in it can help you get a better sense of the scale used.

1 Make a scale drawing of the perimeter of the classroom. On the drawing make an "X" to show your location in the classroom.

2 Draw a scale version of a teacher's desk and a student's desk in the scale 1 cm :1 m.

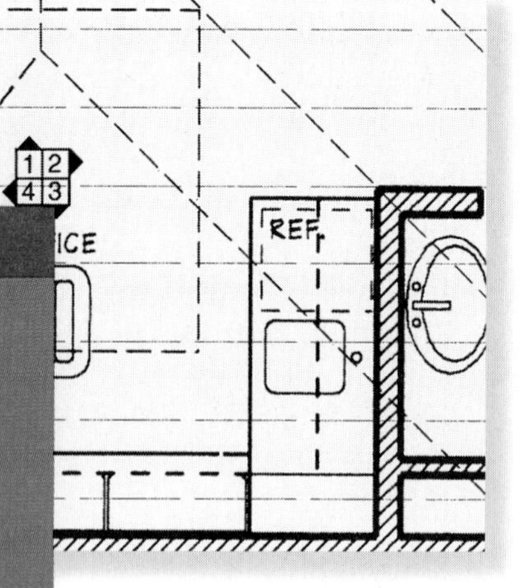

Guidelines for Floor Plans

- The home must be one story.
- Use a scale of one centimeter represents one meter to draw the floor plan and to build the walls and roof.
- The floor space must fit within a 16 meter by 16 meter square.
- The floor plan should *not* be rectangular.

Create a Floor Plan

Design a floor plan that meets the Guidelines for Floor Plans.

- Follow the Guidelines for Floor Plans on page 140.
- Include at least three pieces of furniture on your floor plan. Draw the furniture to scale to give people a better idea of what it might be like to live in this house.

Describe the Floor Plan

Write a description of your floor plan and what it shows.

- How did you decide how large and what shape to make the different rooms?
- How did you decide what size to make the scale drawings of the furniture?
- How can you check a floor plan to make sure that the spaces and sizes of things are reasonable for real people? (Is the bedroom large enough for the bed?)

Reflect on the math you used to design your floor plan.

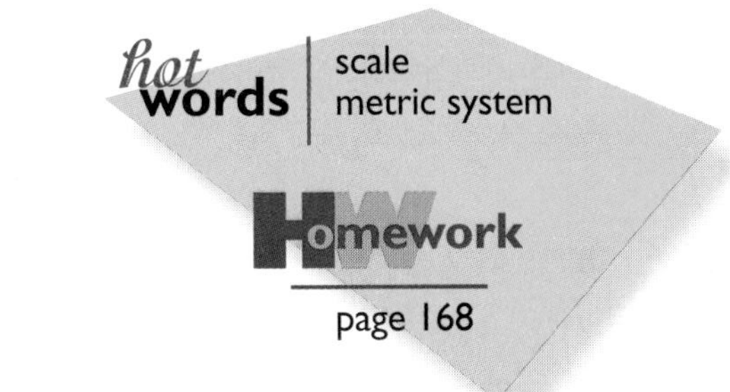

Site Plans and Scale Drawings

USING PROPORTIONAL RELATIONSHIPS IN SCALE DRAWINGS

A site plan shows where the house will be placed on the building site. After placing your floor plan on the scale-size site to make a site plan, you will experiment with scale drawings and add some of them to your building site.

Make a Site Plan

How can you use what you know about scale to make a site plan?

To make a site plan, you will need your floor plan and an $8\frac{1}{2}$ in. by 11 in. sheet of paper that represents the scale size of your building site. Cut out your floor plan and glue it onto your building site. Make sure you follow Building Regulation #105 below.

- How large is your actual site in meters?
- How can you figure out the actual size of your site?

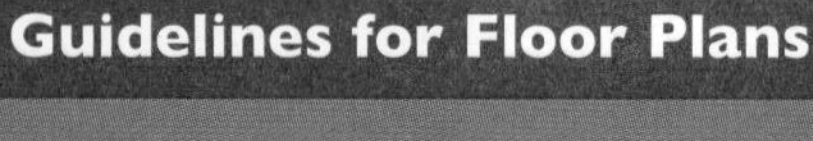

Guidelines for Floor Plans

Building Regulation #105:
The front of the home must be set back at least 5 meters from the street. The rest of the home may be no closer than 2 meters to the property line.

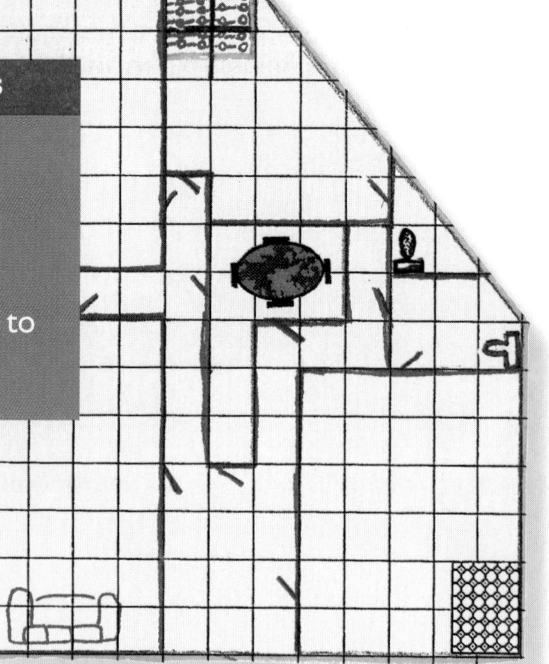

Experiment with Different Scales

How does changing the scale affect the size of a scale drawing?

Use the four different scales shown to make scale drawings of yourself or of another object, such as a tree, that is taller than you.

1. Show the front view of the person or object. (This is different from the top-down view you used in your floor plan.) You do *not* need to show details.
2. Cut out each drawing and label it on the back with the scale you used.
3. After you make the drawings, put them in order by size.
4. Glue the drawing you made in the scale of 1 cm:1 m to a card stock backing. Add tabs to the bottom so you can stand the drawing upright on your site plan.

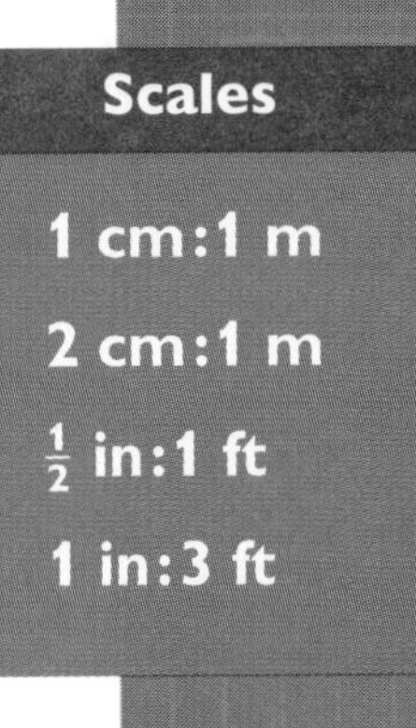

Scales
1 cm:1 m
2 cm:1 m
$\frac{1}{2}$ in:1 ft
1 in:3 ft

Write What You've Learned About Scale

Write your responses to the following questions.

- Suppose you made many copies of the 1 cm:1 m scale version of yourself and then stacked the copies from head to toe. How many copies would you need to reach your actual height?
- What tips would you give to help someone figure out how large an object would be in different scales?
- Why do floor plans and site plans need to be drawn in the same scale?

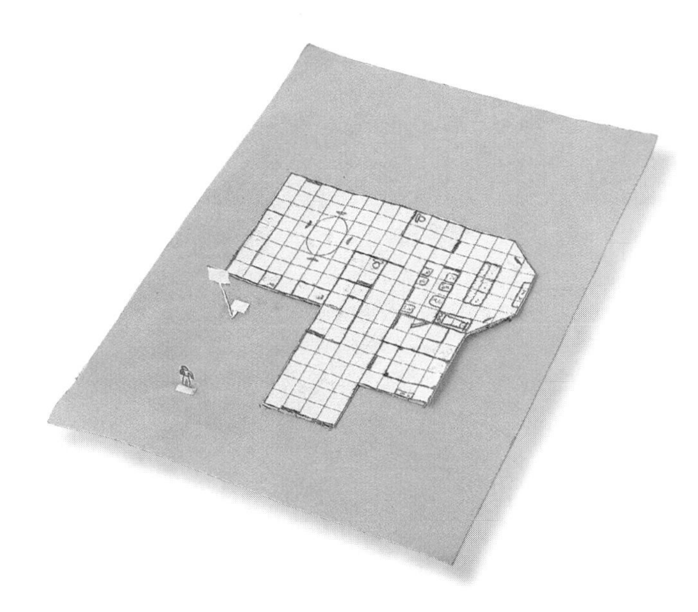

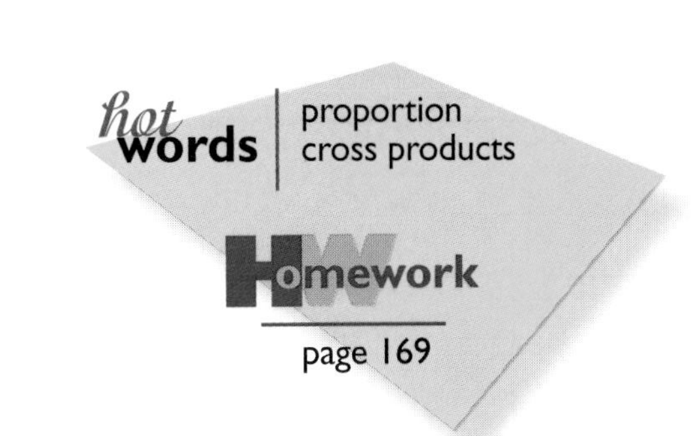

Building the Outside Walls

RELATING ACTUAL SIZE TO SCALE REPRESENTATIONS

Choosing a realistic scale height is an important part of building the walls for your scale model home. To complete the walls, you will also need to think about how to make them match the perimeter of your floor plan.

How can you build outside walls that are a realistic height and match the size and shape of your floor plan?

Construct the Outside Walls

After deciding on your wall height, construct the outside walls around your floor plan. Be sure to follow the Guidelines for Walls.

- Experiment with methods for making the walls. Share methods that work well with your classmates.
- Keep in mind that there must be an exact fit between the outside walls and the perimeter of the floor plan.

Guidelines for Walls

- All outside walls of the house need to be the same height.
- All houses are to be one story.
- All houses must have a front door and back door for fire safety.
- Walls should be made with a scale of 1cm:1m.

Add Windows and Doors to the Walls

Now you are ready to add windows and doors to the walls. You will need to choose a reasonable size for the windows and doors on your scale model.

- Use your own method to add the windows and doors to your walls.
- Be sure to follow the guideline for doors in the Guidelines for Walls.

How can you use scale to decide how high up to place the windows?

How large should doors and windows be on the scale model?

Write About Walls, Windows, and Doors

Write about the thinking you used to make the walls, windows, and doors.

- What height did you choose for your outside walls? Explain why you chose that height.
- What suggestions would you give other students to help them construct walls for a model home?
- How did you decide what size to make the windows and how high to place them?

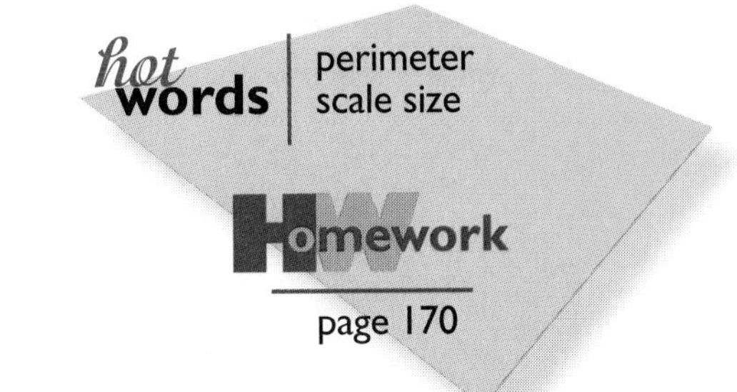

Area and Cost of Walls

FINDING THE AREAS OF RECTANGLES TO ESTIMATE COST

Estimating costs is an important part of any building project. After determining the area of walls for your model, you will be able to estimate the cost of building the actual-size walls. First, can you figure out the cost of a classroom wall?

What would it cost to build the actual walls of your model house?

Estimate the Cost of the Outside Walls

Use your house model to estimate the cost of building materials for the walls of the actual home.

1. Use whatever method you choose to determine the area and cost of the walls.
2. Record the cost on the Cost Estimate Form.

Cost Estimate Form

COST ESTIMATE FORM

	Model Area	Actual Area	Cost/m²	Total Cost	Estimate Checked by
Outside Walls	202.8 cm²	202.8 m²	$53.00	$10,748.40	
Roof					
Ground Floor					
Ceiling					
		Materials Subtotal			
		Labor at 60% of Materials			
		Total Cost Estimate			

one By: ______

Signature ______

hecked By: ______

nt's Signature ______

Cost of Building Materials for Walls

The cost of building materials for walls is $53 per square meter.

(For cost estimates, consider windows and doors to be part of wall area.)

Experiment with Wall Design and Costs

Refer to the Specifications for the Evergreen Project to create a scale drawing of a design for the front wall on centimeter grid paper. Use a scale of 1 cm:1 m. Show the size and position of the door and windows.

1 Use the rate of $53 per square meter to estimate the cost of the wall materials for the design you created. Consider doors and windows to be part of the wall area. Show your work.

2 Cut the cost of the front wall you designed by at least $350, but not more than $450. You can change the length and/or height of the wall, but not the $53 per square meter cost for materials. Describe your solution for cutting costs. Explain why your solution works.

3 Suppose you need to make a larger scale drawing of your wall design. The larger scale drawing should take up most of a 21.5 cm by 55 cm sheet of paper, and cannot be less than 50 cm wide.

a. What scale should you use for the drawing? Explain your thinking.

b. What are the dimensions of the wall in that scale?

Specifications for Evergreen Project

Measurements for Front Wall

Front wall: 4.5 m high by 13 m wide
Front door: 2.3 m tall by 1 m wide
Medium window: 1.5 m high by 1.2 m wide
Large window: 2 m high by 2.5 m wide

Guidelines for Walls

The wall needs to have a front door and 3 windows. Use both medium and large windows.

How can you apply what you have learned about scale, area, and cost relationships?

PHASE TWO

To: Apprentice Architects

Your next task is to make these four roof models: gable, hip, pyramid with rectangular base, and pyramid hip with pentagonal base of unequal sides.

You will design two-dimensional nets that will fold up to make each roof model. Nets aren't used for building actual roofs, but they are a great modeling technique. Each base should be about the size of your palm.

You will design a roof to match the floor plan of your model home. Then you will figure out the cost of building materials for an actual-size version of your model roof.

No house is complete without a roof. In this phase you will experiment with making nets for roof shapes and use a compass to design and construct a roof for your model.

To estimate the cost for roof materials, you will learn to find the area of the shapes that make up the roof. What shapes do you see in the roof nets pictured on these pages? What are the shapes of roofs in your neighborhood?

Roofs, Area, and Cost

WHAT'S THE MATH?

Investigations in this section focus on:

TWO- and THREE-DIMENSIONAL GEOMETRY

- Inventing and adapting techniques for constructing nets that will fold up into prisms and pyramids
- Constructing a triangle, given the lengths of its sides
- Using the compass to measure lengths and construct triangles

ESTIMATION and COMPUTATION

- Using a rate to make cost estimates based on area

AREA

- Building an understanding of area
- Finding the areas of triangles and trapezoids

MathScape Online
mathscape2.com/self_check_quiz

5 Beginning Roof Construction

CREATING NETS FOR THREE-DIMENSIONAL SHAPES

Now you will experiment with making models of four different house roofs. To make each roof model, you will need to design a two-dimensional net that will fold up to make the roof shape.

What are the two-dimensional shapes that make up three-dimensional roof shapes?

Identify the Shapes of Roofs

Describe each roof in geometric terms. For each roof, identify the following:

- What is the three-dimensional shape of the roof?
- What are the two-dimensional shapes that make up the three-dimensional shape?

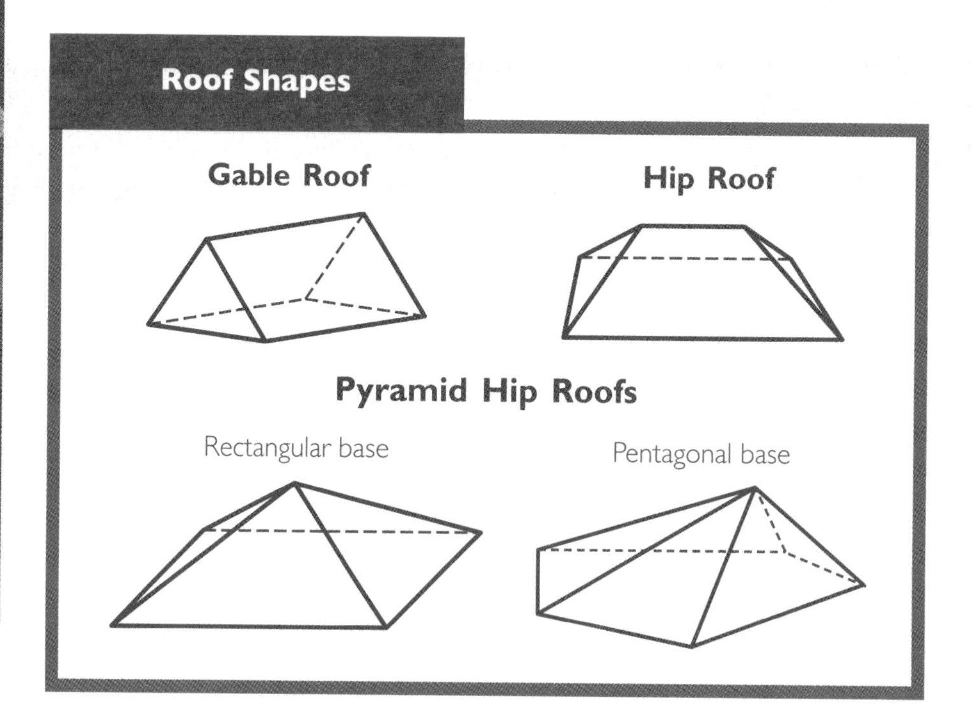

Experiment with Roof Models

Design a net that you can fold up to make a model of one of the four roofs. Cut out the net and fold it into shape.

- The base of the roof should be about the size of your palm.
- As you explore your own construction techniques, try to discover ways to get the parts of the net to match up so that the roof does not have any gaps.

How can you design two-dimensional nets to fold up into three-dimensional roofs?

Write About Construction Techniques

After you finish making a roof, use these questions to write about your experience:

- What was the most challenging part of designing your net?
- What tips would you give someone else?
- What are some differences between two-dimensional and three-dimensional shapes?

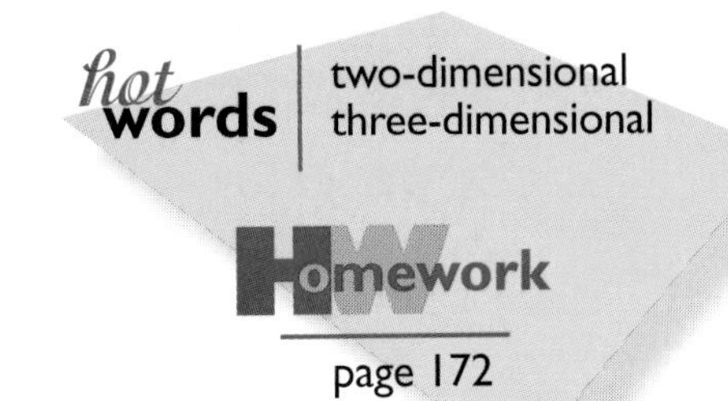

Advanced Roof Construction

CONSTRUCTING TRIANGLES WITH COMPASS AND RULER

The compass techniques on these pages will help you solve many of the problems you discovered as you explored roof construction. Use these techniques to make models of the four roof shapes shown on page 150.

How can you use the compass to make constructions more precise?

Construct Triangles for Roof Nets

You can use a compass and ruler to construct triangles for a net. The Building Tips below illustrate how to construct triangles for a hip roof net. After you construct the triangle for one side of the net, repeat the same steps to construct the triangle for the other side. Use the same method to make a gable roof net.

Building Tips

HOW TO USE A COMPASS TO CONSTRUCT TRIANGLES FOR A HIP ROOF NET

1. Measure edge *CD* by stretching the compass from vertex *C* to vertex *D*.
2. Keep compass at this setting and mark off an arc from vertex *C* and another from vertex *B*.
3. Where arcs cross is the vertex of the triangular face. Use a ruler to draw in the sides of the triangle.

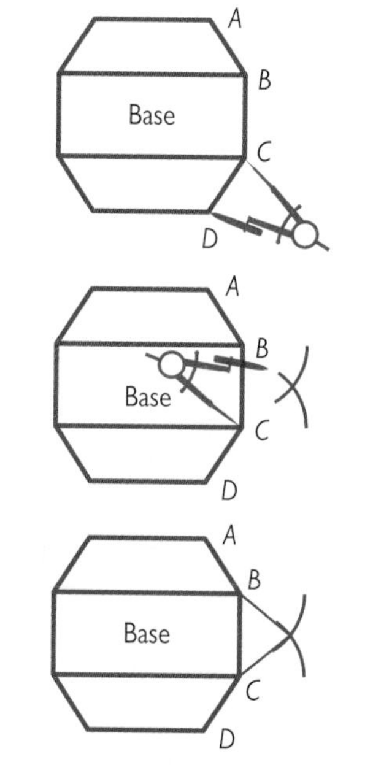

Construct a Pyramid Roof

How can you use a compass to construct a pyramid roof?

You can use the following technique to make a pyramid roof over any polygon—rectangular or nonrectangular. Can you figure out why it works?

Building Tips

HOW TO USE A COMPASS TO MAKE A PYRAMID ROOF

1. Draw the roof base and cut it out. Cut a straw to the roof height you want. Tape the straw vertically at about the center of the roof base.

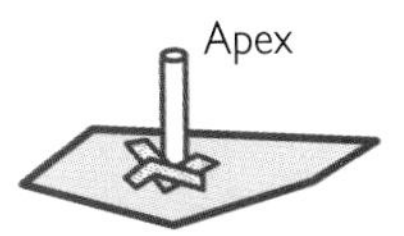

2. Tape a piece of paper onto one edge of the base. Put the point of the compass at a vertex (*B*) of the roof base and stretch the compass to the top of the straw.

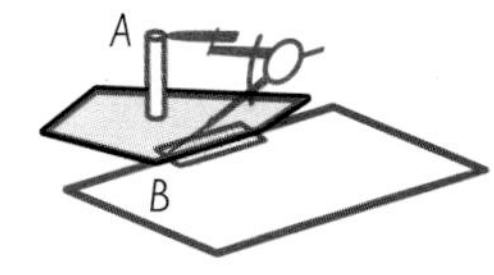

3. Keeping the point of the compass on *B*, draw an arc on the paper. Repeat Step 2 with the other vertex (*C*), and draw another arc.

4. Use a ruler to draw lines connecting *B* and *C* to the point *D* where the arcs cross. This triangle is one roof panel.

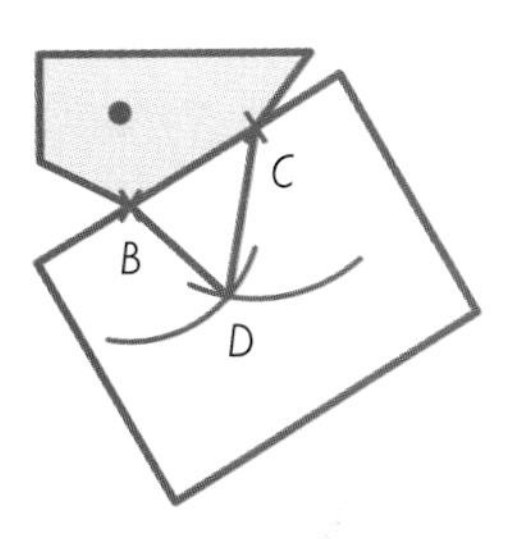

5. Cut out the roof panel. If you leave a tab on one edge, your roof will be easier to tape together.

6. Continue to measure and lay out the rest of the roof panels. Remove the straw before you fold up and tape the roof.

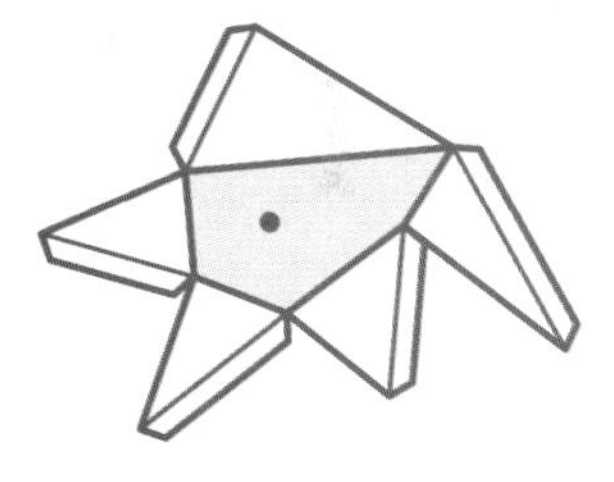

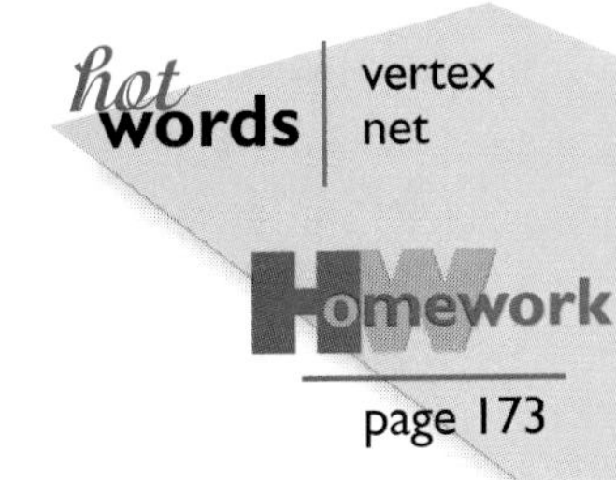

Determining Roof Area

FINDING AREAS OF TRIANGLES AND TRAPEZOIDS

To find the cost of making your house model roof, you will need to find the areas of the shapes that make up the roof. After learning to find the area of a triangle by four different methods, you will consider how to find the area for a trapezoid.

Explore Triangle Area Methods

Why do different methods for finding the area of a triangle work?

On the handout Incomplete Net for a Hip Roof, experiment with each of the four different methods to find the area of the same triangular panel of the net. Write about why you think each method works.

Methods for Finding Triangle Area

Cover and Count

Cover the triangle with a transparent grid of centimeter squares. Count the number of whole squares plus the number of squares that can be made out of the leftover pieces.

Surround with a Rectangle

Draw a rectangle that surrounds the triangle and touches its vertices. Measure the length and width of the rectangle. Use the formula $A = lw$ to find its area. Divide the area by 2 to get the area of the triangle.

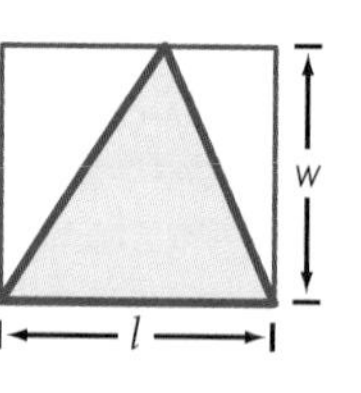

The Formula $A = \frac{1}{2}bh$

Draw in the altitude at a right angle to the base. Measure the base (b) and the height of the altitude (h). Use the formula $A = \frac{1}{2}bh$ to calculate the area.

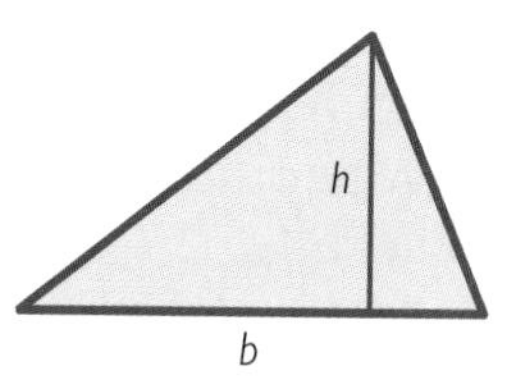

Draw the Midpoints Rectangle

Draw the midpoints rectangle. Measure its length and width. Use the formula $A = lw$ to find its area. Multiply the area by 2 to get the area of the triangle.

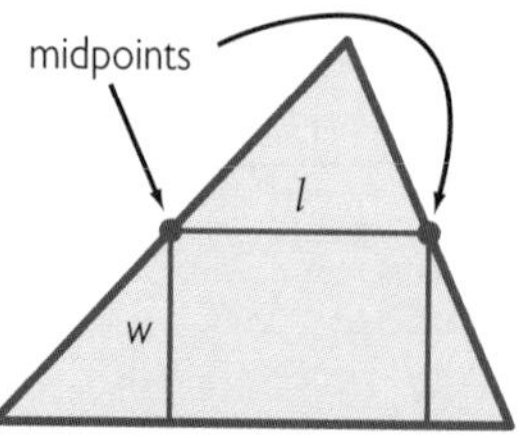

Find the Area of a Trapezoid

Look at the trapezoidal roof panels on the handout Incomplete Net for a Hip Roof. Which of the methods you used to find triangle area could you use to find the area of a trapezoid?

1 Find the area of the trapezoid in at least two different ways. Record the methods you used and the results. Explain why each method works.

2 Determine the total area of your hip roof model. Do not include the area of the roof base.

What methods can you use to find the area of a trapezoid?

Hint

FINDING THE FORMULA FOR THE AREA OF A TRAPEZOID

1. Label the trapezoid cutout as shown below.

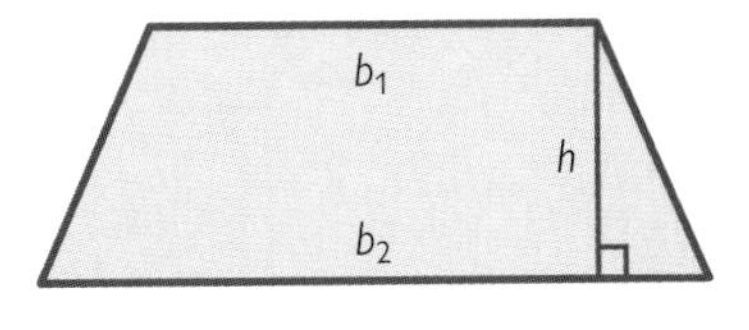

2. Use a ruler to draw a diagonal line and cut the trapezoid into two triangles, as shown.

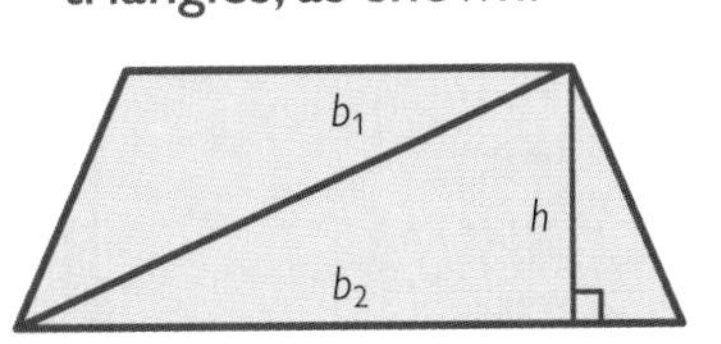

3. Can you arrange the two triangles to form one large triangle? Use what you know about the area of the large triangle to come up with a formula for the area of a trapezoid.

Calculating Roof Cost

ESTIMATING COST BASED ON AREA

Now you are ready to apply what you've learned about roof construction to make a roof for your model home. After estimating the area and cost of the roof, you will attach it to your model.

How can you design a roof base, choose a roof style, and construct a roof for your model home?

Design and Construct a Roof

Read the Guidelines for Roofs below.

1. Design the roof base and choose a roof style.
2. Use the techniques you learned on pages 150–153 to construct a net for the roof of your house model.

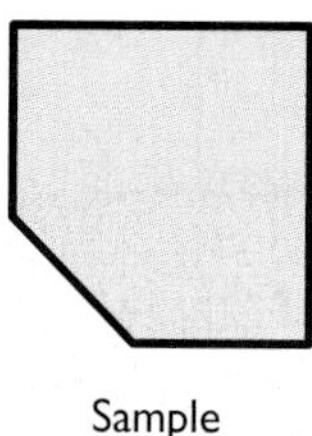

Sample Floor Plan

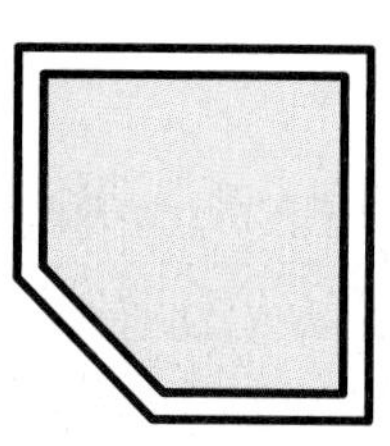

Roof Base for Pyramid Hip Roof

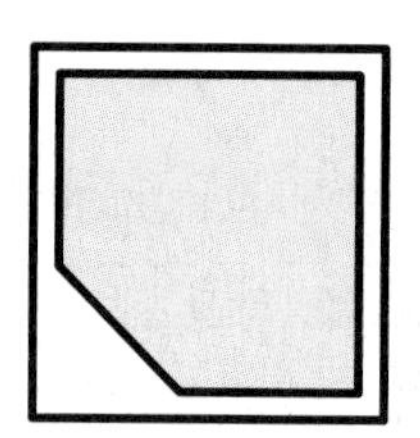

Rectangular Roof Base

Guidelines For Roofs

All roofs must be sloped, not flat.

The roof base must overhang the floor plan by at least half a meter so that rain coming off the roof will drip down away from the walls.

If the perimeter of the floor plan is concave on any side, the roof base must be drawn convex to overhang that area. This will make it easier to construct the roof.

Find the Area and Cost of the Roof

Figure out the cost of the materials for your roof, based on the amount in the Cost of Roof Materials. Remember that the roof base should not be included in the roof area.

1. Draw a sketch of your roof and label each panel with a different letter (A, B, C, D, and so on). Find the cost of each panel. Record your work and label it with the panel letter.
2. Write about the method you used for finding area and why you chose to use that method.
3. On your Cost Estimate Form, fill in the model area, actual area, cost per square meter, and total roof cost.

How can you apply what you know about area to find the area and cost of the roof?

Cost of Roof Materials

The cost of roof materials is $48 per square meter.

Rate Your Model Home

Design

- Is the roof base a good fit for the shape of the floor plan?
- Is the shape of the roof a good match for the home?
- On a scale of 0 to 10, with 0 meaning "dull" and 10 meaning "innovative," how would you rate your home? Why?

Construction

- Do all sides of the roof match up precisely?
- Does the overhang follow the Design Guidelines?
- On a scale of 0 to 10, with 0 meaning "poorly crafted" and 10 meaning "masterpiece," how would you rate your home? Why?

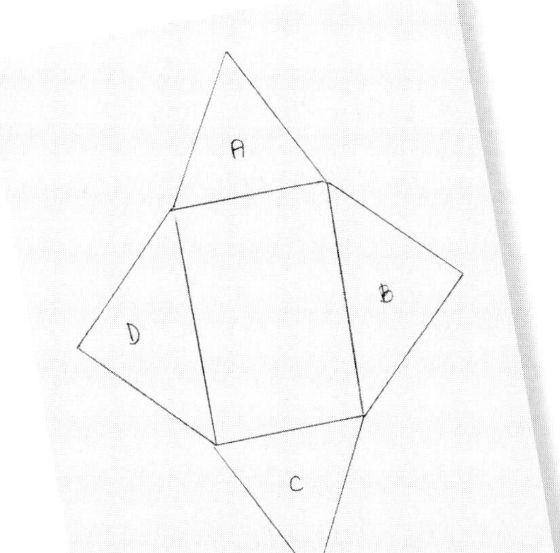

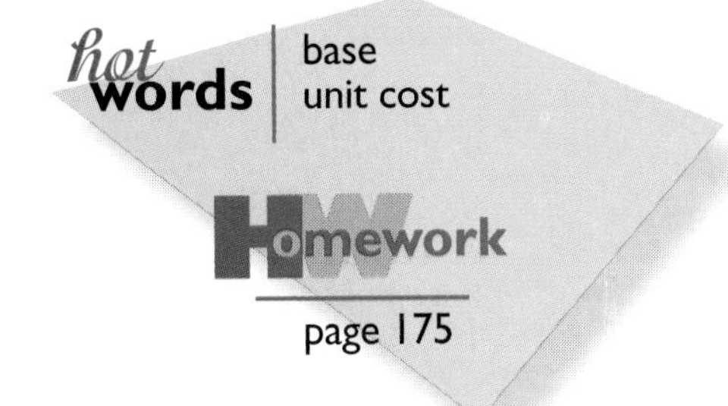

PHASE THREE

Budgeting is an important part of building. To complete the cost estimate for your model home, you will need to find the area and cost of your floor and ceiling. Then you will need to add on the cost of labor.

By comparing and analyzing the costs of the different model homes completed in the last phase, you will learn some things about budgeting and building. This will prepare you for the final project: designing a new home, within a given budget, for a climate and building site of your choice.

Budgeting and Building

WHAT'S THE MATH?

Investigations in this section focus on:

SCALE and PROPORTION

- Making accurate drawings and scale models

GEOMETRY

- Planning and constructing three-dimensional models

AREA

- Applying methods for finding the areas of rectangles and triangles to finding the area of any polygon
- Planning and revising designs to fit within a given range of area measurements

ESTIMATION and COMPUTATION

- Planning and revising designs to fit within a given range of costs

MathScape Online
mathscape2.com/self_check_quiz

Costs of Floors and Ceilings

FINDING THE AREAS OF POLYGONS TO ESTIMATE COST

Your next step is to estimate the materials cost for the floor and ceiling of your house. The floor plans and ceilings fit into the geometric category of *polygons*. You will experiment with methods for finding the area for any polygon so that you can estimate the cost of your own floor and ceiling.

How can you find the area of any polygon?

Investigate Polygon Area Methods

Figuring out the area and cost of your floor plan and ceiling is a challenging task. Since they have irregular shapes, there's not one formula you can use for the areas. However, there are many different strategies that will work. See what you can come up with by experimenting with a sample floor plan.

1. What are the area and cost of the sample floor plan on the handout Sample Floor? Show your work. Keep notes about how you found the answers so you can report back to the class.
2. Describe how you would figure out the areas of the ceilings shown below.

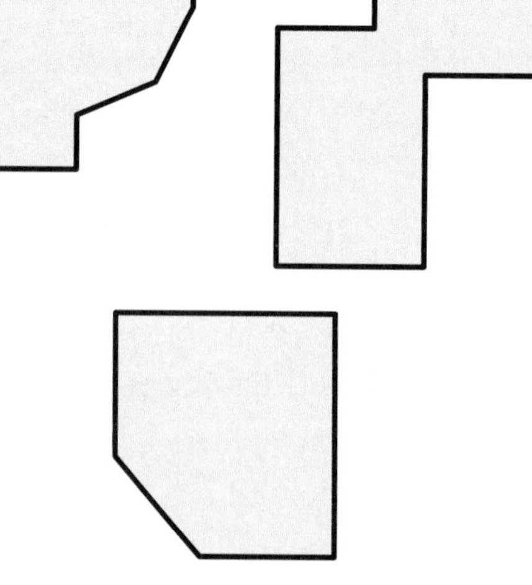

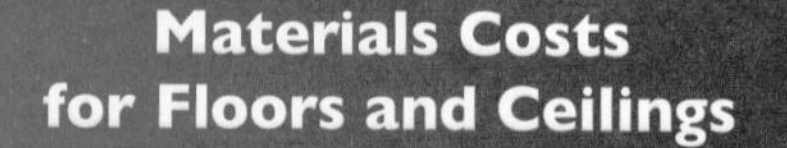

Materials Costs for Floors and Ceilings

- Floor materials cost $95 per square meter.
- Ceiling materials cost $29 per square meter.

Find the Floor and Ceiling Materials Costs

You will need a copy of your floor plan. To make a ceiling plan, make a copy of your roof base with overhangs.

1. What is the cost of materials for your floor and ceiling? Show your work.
2. Check to be sure the costs are reasonable. Then add the costs to your Cost Estimate Form.

Compare Model Home Areas and Costs

Write about how the costs of the model homes compare.

- How does the area of your floor compare with the area of the ceiling, walls, and roof?
- How do the costs of your floor and ceiling compare with your classmates' costs?
- What are some ways to determine whether or not a cost estimate is reasonable?

Cost Estimate Form

COST ESTIMATE FORM

	Model Area	Actual Area	Cost/m²	Total Cost	Estimate Checked by
Outside Walls	202.8 cm²	202.8 m²	$53.00	$10,748.40	
Roof	256 cm²	256 m²	$48.00	$12,288.00	
Ground Floor	151 cm²	151 m²	$95.00	$14,345.00	
Ceiling	179 cm²	179 m²	$29.00	$5,191.00	
		Materials Subtotal			
		Labor at 60% of Materials			
		Total Cost Estimate			

Estimate Done By:
Architect's Signature ______

Estimate Checked By:
Accountant's Signature ______

10 Adding the Cost of Labor

USING PERCENTAGE TO ESTIMATE COST

The final thing you need to consider for your model home is the cost of labor. Labor is people working to put the materials together into an actual house. How do you think the total cost of your model home will compare to the total cost of your classmates' models?

How can you find the cost of labor and estimate the total cost for your home?

Find the Labor Cost and Total Cost Estimate

Read the information on Cost of Labor below.

1. Figure out the cost of labor for your home.
2. Describe how you figured out the cost of labor. Fill in the amount on your Cost Estimate Form.
3. Figure out the total estimated cost for your home. Fill in the amount on your Cost Estimate Form.

Check Cost Calculations

Work with a partner to check each other's Cost Estimate Forms. By helping each other, you can make sure that you both have accurate cost estimates for your model homes.

Explain to your partner how you figured out the costs for your home. Then your partner will take on the role of Accountant and check all your calculations. If the Accountant finds any mistakes, you will need to make corrections on your Cost Estimate Form.

Cost of Labor

Labor costs are an additional 60 cents for every dollar of material costs. This means that labor costs are 60% of material costs.

Compare the Total Costs of the Homes

Copy the class data for all the homes.

1. What is the typical cost for the homes the class has made? (Find the mean and the median for the data.) Make a graph of the data on house costs.
2. What did you find out from the graph? Write a summary of the data.
3. How could you change your home so that it costs the median amount?
4. The bar graph below displays total cost data from the houses built by one class. How does your class's data compare?

How can you compare data on home costs?

Find the Square Unit Cost of Floor Space

Architects and realtors often use the cost per square unit of floor space to compare homes.

- How much does your home cost per square meter of floor space? For example, a $71,000 home with 155 square meters of floor space would cost about $458 per square meter of floor space.
- What information does the cost per square meter of floor space tell you about a house? How might that information be useful for comparing homes?
- How does your cost per square meter of floor space compare with the cost per square meter of your classmates' homes?

Cost of Model Homes

Cost: 100,000; 90,000; 80,000; 70,000; 60,000; 50,000; 40,000; 30,000

Number of homes: 0; 1; 2; 3; 4

hot words: mean, median

page 177

Designing Within a Budget

FINAL PROJECT

What have you learned about scale, geometry, area, measurement, and cost calculation? As a final project, you will apply what you have learned to design the best home you can. This time you will be designing for a certain climate and budget.

How will you design a home so that its cost is close to the target cost?

Make a Plan for the Final Project

Read the Final Project Design Guidelines. Choose one of the three building sites shown on this page and one of the four climate options on the handout Climate Options. Think about how the features of the land and the climate you chose will affect your home design.

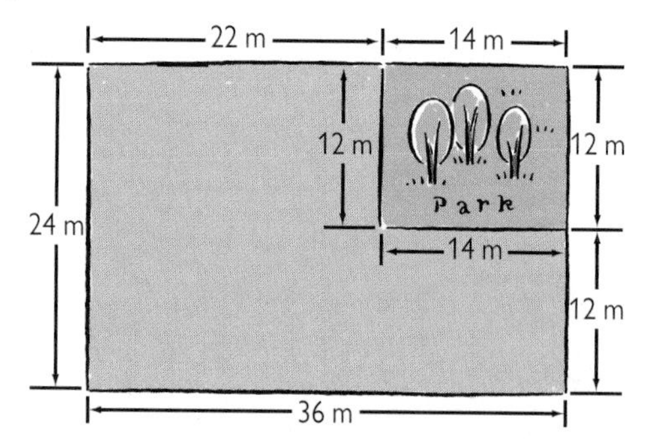

1. Make a scale drawing of your lot, using a 1 cm:1 m scale.

2. Write a brief description of your home.

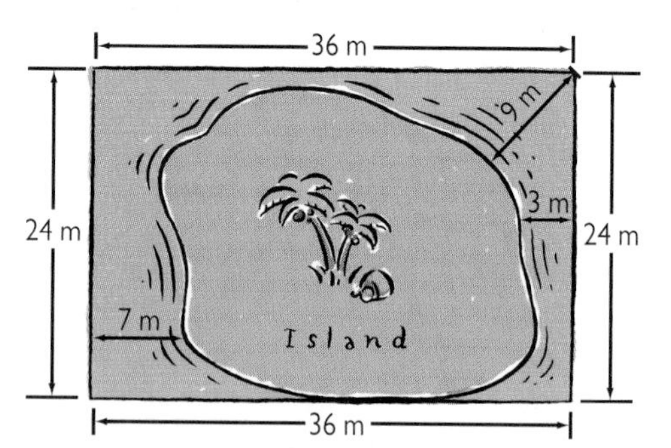

Final Project Design Guidelines

- The home should be suited to the chosen land and climate.
- The area of the ground floor of the home must be at least 150 square meters.
- The cost of building the home must be close to the target amount for the chosen climate.

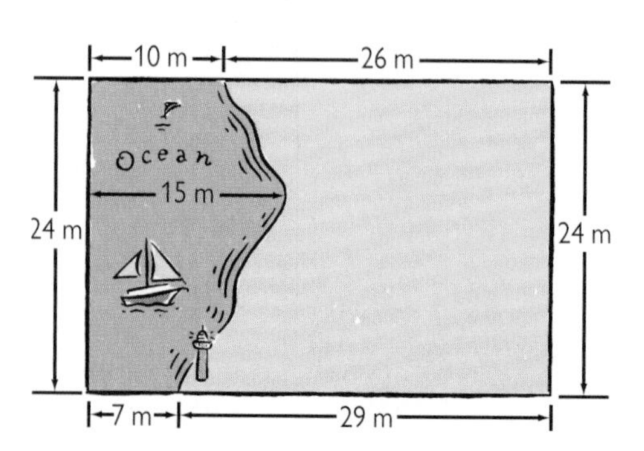

Work Independently on the Final Project

How can you use what you've learned to design the best home you can?

Use what you have learned about constructing floor plans, walls, and roofs, and about estimating costs to design the best home you can. Be sure to follow the Final Project Design Guidelines on page 164.

1. **Design a floor plan.** Draw a floor plan for the home using a scale of 1 cm:1 m. The home may have more than one story.
2. **Make a rough estimate.** Try to figure out if the cost of your home will be close to, but not go over, the target cost for your climate. If necessary, change your design.
3. **Build a model.** Use a scale of 1 cm:1 m. Make a carefully constructed model out of paper. Attach it to the lot.
4. **Determine the cost.** Figure out the cost of building your home. Show all the calculations and measurements you used in making your cost estimate.

Write About the Home

When you have finished building your home, write a sales brochure or a magazine article describing it. Include the information described on the handout All About Your Home.

Touring the Model Homes

COMPARING AND EVALUATING HOME DESIGNS

It can be fun to see the different homes other people have designed for the same climate and budget as yours. How would you rate your final project? After you complete the House-at-a-Glance Cover Sheet and rate your own project, you will compare the homes.

How do the house designs, descriptions, and costs compare?

Compare the Houses

Make comparisons among the different houses your classmates have designed.

- What are the similarities and differences among homes for a particular climate?
- For a particular climate, how do homes with a low cost per square meter of living space compare with those with a high cost per square meter?
- What features do homes designed for similar building sites share?
- What different roof styles are used?

Reflect on the Project

- What do you like best about your home? What would you do differently if you designed another home?
- What mathematics did you use to design and build your house, figure its cost, and write your sales brochure?
- Suppose someone wants to build your house, but tells you it costs $10,000 too much. Describe how you might change your design to reduce the cost.
- Choose a different climate. What changes would you make to adapt your home to that climate?
- What does your project show about what you learned during this unit? What are some things you learned that your project doesn't show?
- Write a description of this unit for students who will use it next year. What math will they learn?

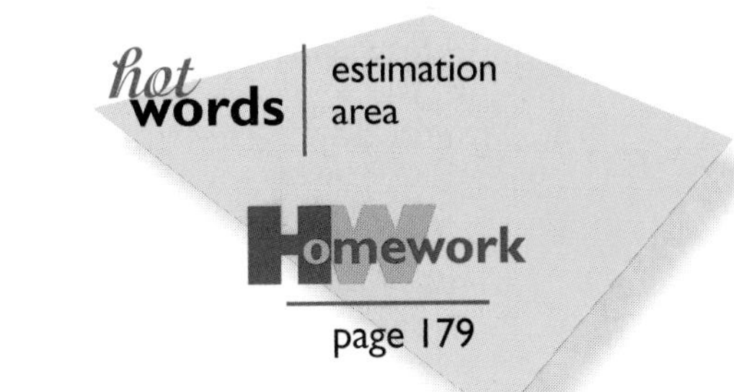

Homework 1

Designing a Floor Plan

Applying Skills

The scale drawings below are shown in a scale of 1 cm:1 m. Give the actual dimensions of each object. To do this you will need a metric ruler to measure the scale drawings.

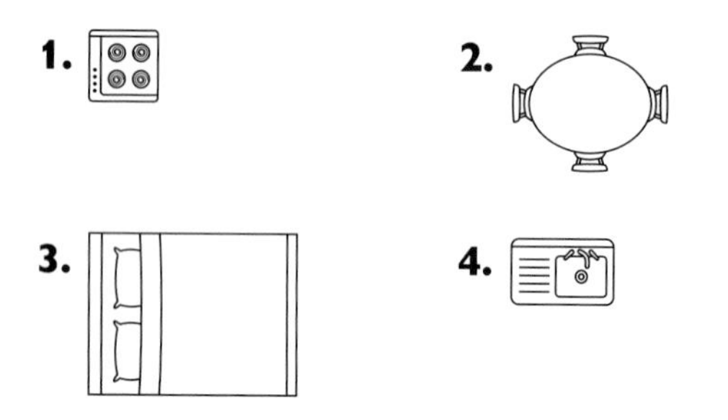

Find the scaled-down dimensions of each room on a floor plan with a scale of 1 cm:1 m.

5. a bedroom that is 4 meters long and 3 meters wide

6. a kitchen that is 3.5 meters long and 2.8 meters wide

7. a living room that is 4.5 meters long and 5.75 meters wide

Tell what each total measurement would be in meters.

8. 23 centimeters

9. 80 centimeters

10. 5 centimeters

11. 1 meter and 15 centimeters

12. 2 meters and 85 centimeters

13. 1 meter and 8 centimeters

Extending Concepts

Make a scale drawing of each object, using an inch ruler and a scale of 1 in:1 ft. Show the top-down view.

14. a bathtub that is 4 ft 10 in. by 2 ft 4 in.

15. a piano with a base 4 ft 2 in. by 22 in.

16. a dresser with a base 31 in. by 18 in.

Find the dimensions of each room if it was drawn on a floor plan using a scale of 1 in:1 ft.

17. 12 ft 6 in. by 11 ft 6 in.

18. 13 ft 3 in. by 12 ft 9 in.

19. Estimate the length and width of your bedroom. Then find its dimensions in a scale of 1 in:1 ft.

Writing

20. Answer the Dr. Math letter.

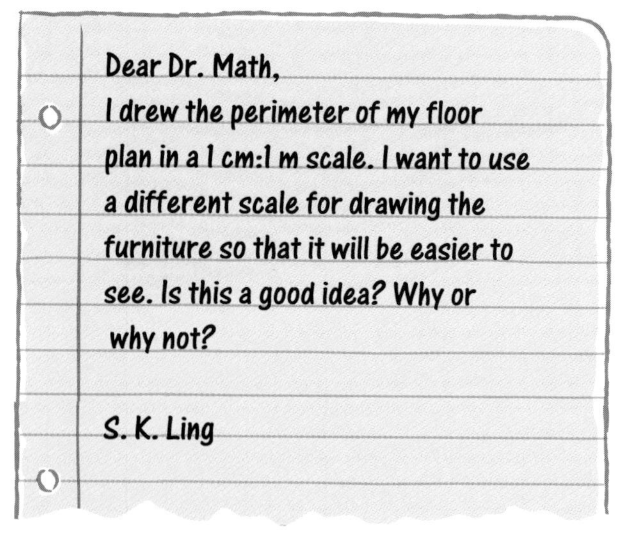

Dear Dr. Math,
I drew the perimeter of my floor plan in a 1 cm:1 m scale. I want to use a different scale for drawing the furniture so that it will be easier to see. Is this a good idea? Why or why not?

S. K. Ling

Site Plans and Scale Drawings

Applying Skills

Give the dimensions of each object using the scale indicated.

	Actual-Size Object	Scale
1.	1 ft by 3 ft	2 in:1 ft
2.	2 ft by 4 ft	2 in:1 ft
3.	3.5 ft by 5 ft	2 in:1 ft
4.	4 m by 3 m	2.5 cm:1 m
5.	1.5 m by 2.5 m	2.5 cm:1 m
6.	5.7 m by 13 m	2.5 cm:1 m

Give the height of a 5-meter-tall tree in each scale.

7. 3 cm:1 m

8. 1.5 cm:1 m

9. 1 cm:2 m

A basketball hoop is 10 ft tall. Give its height in each scale.

10. 1 in:3 ft **11.** 2 in:1 ft **12.** 1.2 in:1 ft

For items 13–14, refer to the scales below.

4 cm:1 m 1 cm:4 m 1 cm:2.5 cm

13. Of the scales listed above, which scale would result in the largest scale drawing of a chimney that is 9 m tall?

14. Which of the scales above would result in the smallest scale drawing of a chimney that is 9 m tall?

Extending Concepts

15. Measure or estimate the dimensions of two objects in your room. Write down their dimensions. What would the dimensions of each object be in a scale of 2 cm:1 m?

16. An apartment is 36 ft long and 25 ft wide. Give two scales you could use to make a sketch of the apartment floor plan that would fit on an $8\frac{1}{2}$ in. by 11 in. sheet of paper and take up at least half of the paper.

Making Connections

17. On average, Earth is 93 million miles from the sun. This distance is called one astronomical unit (A.U.). On a sheet of notebook paper, sketch a scale drawing that shows the distance of each planet from the sun, with the sun at one edge and Pluto at the other.

Planet	Average Distance from the Sun (A.U.)
Mercury	0.4
Venus	0.7
Earth	1.0
Mars	1.5
Jupiter	5.2
Saturn	9.5
Uranus	19.2
Neptune	30.1
Pluto	39.4

Building the Outside Walls

Applying Skills

Find the perimeter of each floor plan shown below. Assume that each square of the grid measures 1 m by 1 m.

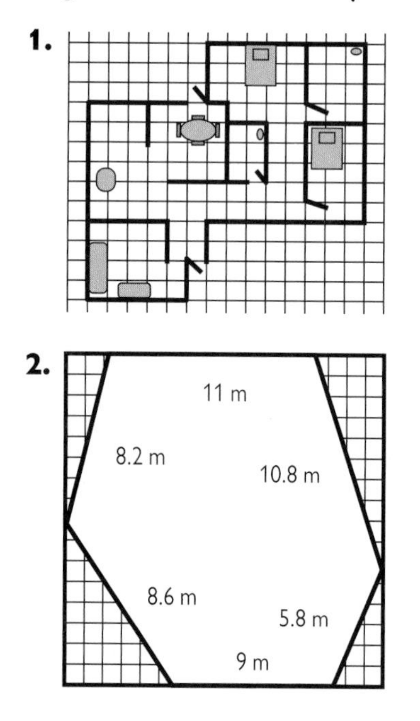

A sliding glass door is 84 inches tall and 72 inches wide. Give its dimensions on a wall plan using each of the following scales.

3. $\frac{1}{2}$ inch : 6 inch

4. $1\frac{1}{2}$ inch : 8 inch

5. $\frac{1}{2}$ inch : 1 foot

6. 3 inch : $1\frac{1}{2}$ feet

Extending Concepts

7. The rectangular lot for a model home has a perimeter of 100 cm. List the dimensions of as many rectangles as you can that have this perimeter. Use only whole-number dimensions.

8. For the model home in item 7, give three examples of rectangles with dimensions that are not whole numbers.

9. Suppose the average perimeter of a model house in a class is 75 cm and the wall height for all models is 2.5 cm. Estimate how many pieces of paper measuring 30 cm by 30 cm you would need for a class to build the walls of 32 model houses.

Making Connections

The type of glass used in windows is called *soda-lime glass.* It contains about 72% silica (crushed rock similar to sand), 15% soda, and 5% lime. Use this information to answer items 10 and 11.

10. How much silica, soda, and lime is there in 180 pounds of soda-lime glass?

11. What percentage of soda-lime glass is *not* silica, soda, or lime? Explain your reasoning.

Homework 4

Area and Cost of Walls

Applying Skills

Find the area of each rectangle. Assume that each square on the grid measures 1 cm by 1 cm.

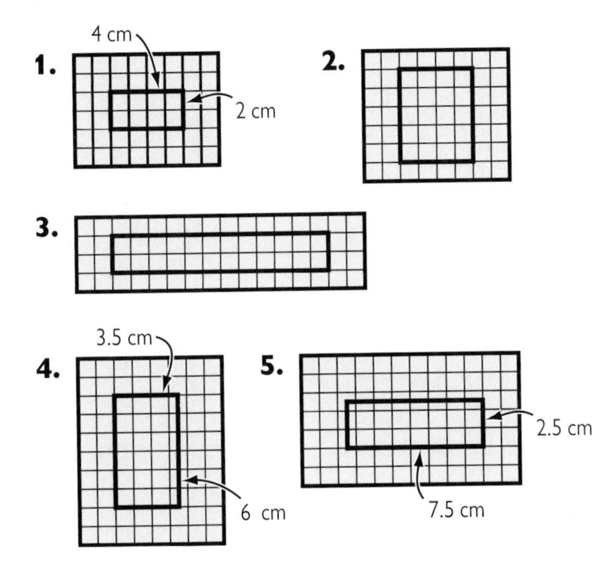

If materials for each square meter of outside wall cost $53, find the materials cost for each of the following.

6. an outside wall 2 m tall and 8 m long

7. an outside wall 2.5 m tall and 6.8 m long

8. the outside walls of a building with a perimeter of 60 m and walls 2 m tall

9. the outside walls of a building with a perimeter of 125 m and walls 2.4 m tall

Extending Concepts

10. Suppose the outside walls of a house are 2 meters high and the amount you can spend for outside wall materials cannot go over $7,500. For each material at the cost given in the table, what is the greatest perimeter the house could have?

Cost of Outside Wall Materials	
wood	$53 per square meter
stucco	$57 per square meter
brick	$65 per square meter
vinyl	$62 per square meter

For items 11–12, suppose you cannot spend more than $13,500 on outside wall materials for a house with a perimeter of 88 m.

11. For outside walls 2.5 m high, which of the outside wall materials listed in the table above could you afford?

12. Find the greatest wall height (to the nearest centimeter) that you could afford for outside walls of brick.

Making Connections

13. The Great Wall of China, dating from the third century B.C., is 2,400 kilometers long and from 6 to 15 meters tall. Estimate the area of one side of the Great Wall. Explain how you made your estimate.

Homework 5: Beginning Roof Construction

Applying Skills

For each figure below, name the three-dimensional shape. Name each two-dimensional shape that would be needed for a net of the three-dimensional shape. Sketch a net for the three-dimensional shape.

1.

2.

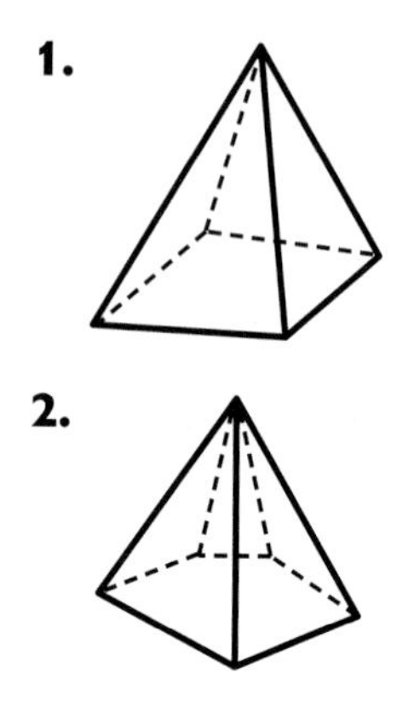

Tell whether each polygon would fold up to form a roof with no gaps. If it would, name the type of roof that would be formed.

3.

4.

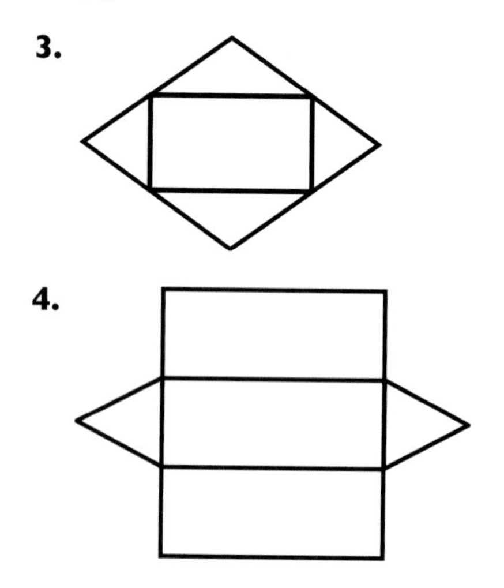

Extending Concepts

For each figure below sketch a net for the three-dimensional shape. Name each two-dimensional shape that appears in the net.

5. **6.**

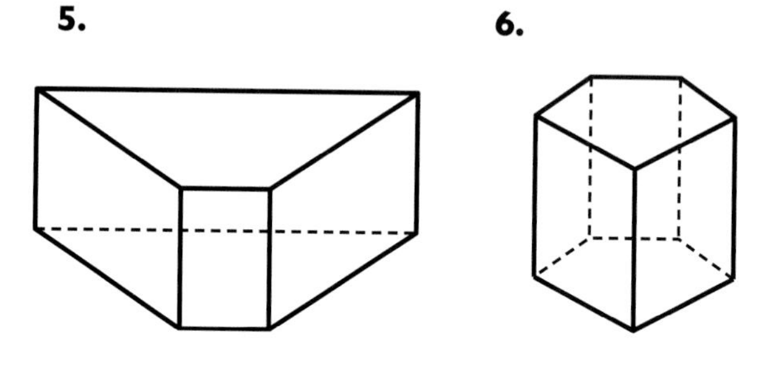

Making Connections

7. The Parthenon is an ancient temple in Greece. Imagine that you are building a scale model of the Parthenon. The width of your model is 32 cm and its length is 70 cm. The height of the gable at each end of the model is 6 cm.

a. Sketch a roof net for your scale model. Give the dimensions of your net.

b. Name the shapes you used.

8. Suppose the height of a column in the Parthenon is 10 m. Make a scale drawing of yourself standing beside the column. Tell the scale you used.

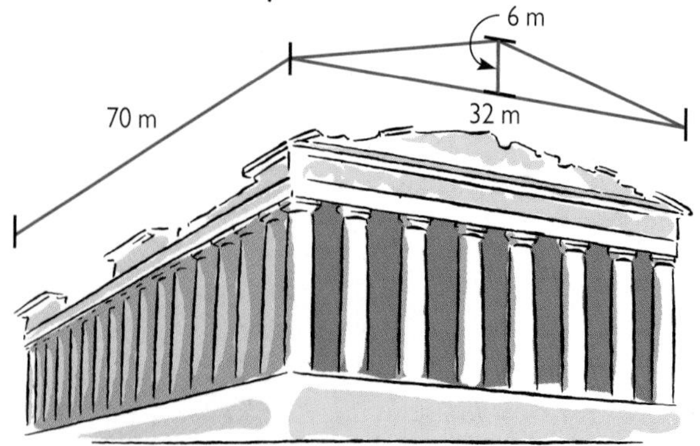

Advanced Roof Construction

Applying Skills

Use a compass and ruler to construct a triangle with the given side lengths.

1. 4 cm, 6 cm, 7 cm

2. 2 in., 2 in., 1 in.

3. 3 cm, 10 cm, 11 cm

4. 8.5 cm, 9 cm, 10 cm

5. 2 in., 2.5 in., 3 in.

6. 5 cm, 6 cm, 10 cm

Tell whether each set of side lengths can or cannot be used to make a triangle.

7. 2, 3, 4

8. 1, 2, 4

9. 11, 12, 14

10. 4.1, 5.5, 9.6

Extending Concepts

The chart gives the lengths of the sides of some triangles. Use what you know about triangles to give a possible length for the third side.

	Length of First Side	Length of Second Side	Length of Third Side
11.	5.3 cm	12 cm	
12.	30 m		40 m
13.		20 cm	30 cm
14.	15.04 m		15.04 m

For items 15–16, suppose that you have three 2-inch sticks, three 3-inch sticks, and three 5-inch sticks.

15. List all of the possible combinations of three sticks that can make a triangle. For instance, two 2-inch sticks and one 3-inch stick will make a triangle.

16. Suppose that you also have three 9-inch sticks. List all triangles that can be made with the 2-inch, 3-inch, 5-inch, and 9-inch sticks.

Making Connections

17. The Great Pyramid of Cheops was built in Egypt about 4,500 years ago.

a. Sketch a net for a model of this pyramid with a square base.

b. Which sides of your net would you measure out with a compass in order to make them the same length? Use an "X" to mark those sides on your net.

Homework 7

Determining Roof Area

Applying Skills

Find the area of each triangle or trapezoid.

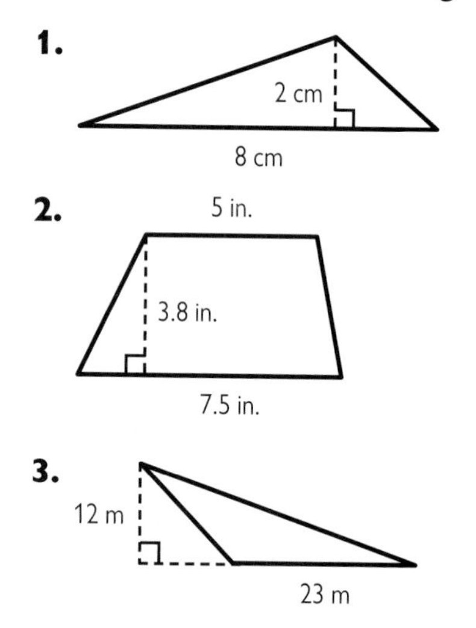

Find the area of each triangle.

4. $b = 5.8$ cm, $h = 16.2$ cm

5. $b = 88.6$ mm, $h = 23.4$ mm

Find the area of each trapezoid.

6. $b_1 = 7.0$ m, $b_2 = 9.7$ m, $h = 8.6$ m

7. $b_1 = 1.1$ m, $b_2 = 3.7$ m, $h = 0.6$ m

8. Find the area of the trapezoid.

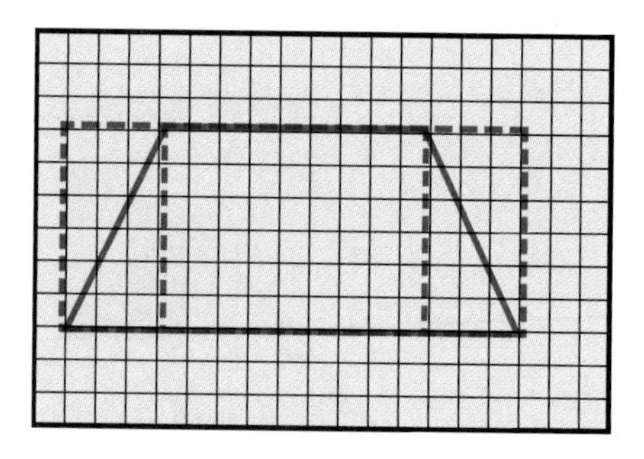

Each square represents one square centimeter.

Extending Concepts

9. In item 8, what is the area of the rectangle that surrounds the trapezoid? What is the area of the rectangle inside the trapezoid? How does the average area of the two rectangles compare to the area of the trapezoid? Will this be true for any trapezoid?

Making Connections

10. a. Find the areas of the triangle and rectangle shown.

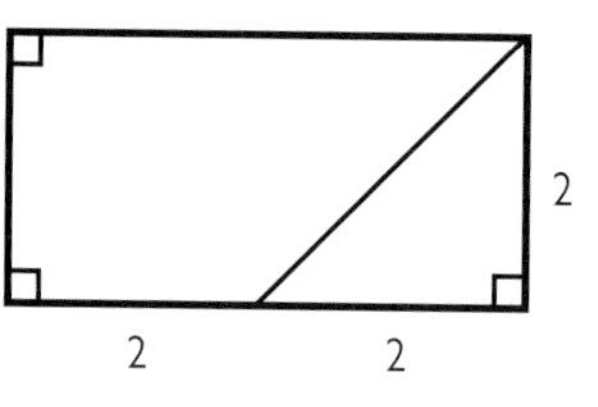

b. What percentage of the area of the rectangle is the area of the triangle?

c. Suppose that the areas of a second triangle and rectangle are in the same proportion and that the area of the rectangle is 40 cm^2. Set up and solve a proportion to find the area of the triangle.

Homework 8: Calculating Roof Cost

Applying Skills

Assume that each triangle or trapezoid below is part of a roof. Find its cost if roofing materials cost \$51 per square meter.

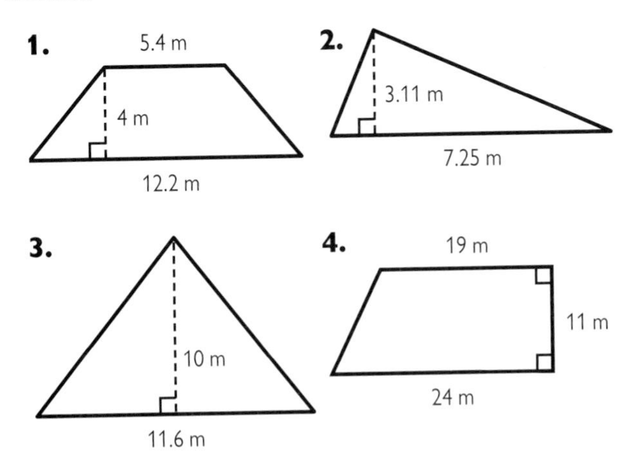

If roofing materials cost \$65 per square meter, find the cost of each roof. You may assume in each case that opposite faces are identical.

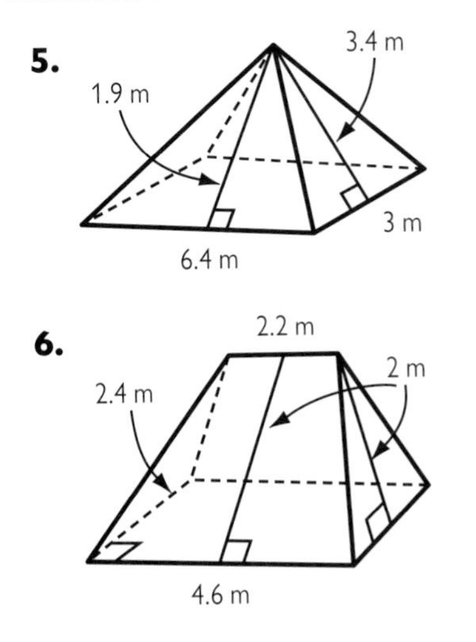

Extending Concepts

7. Two pyramid hip roofs have identical square bases. Each triangular face of the first roof has height 2.2 m. Each triangular face of the second roof has height 6.6 m. If roofing materials for both roofs cost \$48 per square meter, how many times more expensive is the second roof than the first? Explain your reasoning.

8. When architects design a flat roof they must consider the weight of rain water the roof can support. If 1 m^2 on a roof is covered by water 1 cm deep, the water weighs 10 kg. Suppose a flat rectangular roof measuring 10 m by 20 m is covered by water 1 cm deep. How much does the water on the roof weigh?

Writing

9. Answer the Dr. Math letter.

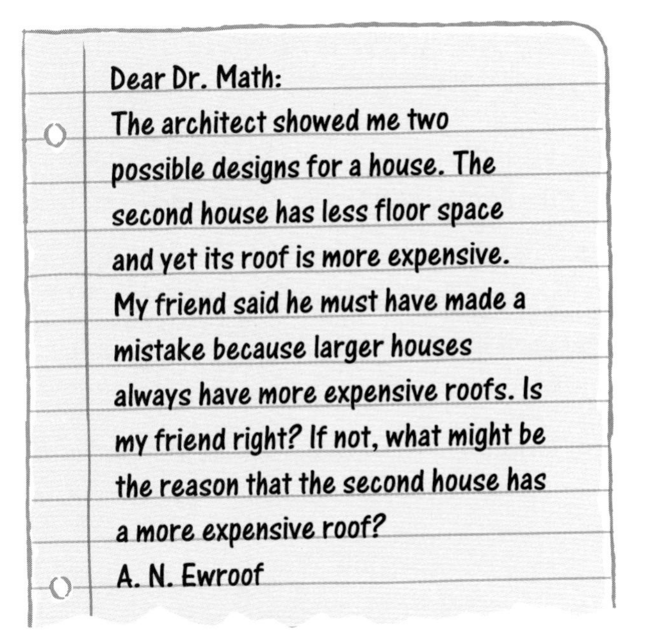
Dear Dr. Math:
The architect showed me two possible designs for a house. The second house has less floor space and yet its roof is more expensive. My friend said he must have made a mistake because larger houses always have more expensive roofs. Is my friend right? If not, what might be the reason that the second house has a more expensive roof?
A. N. Ewroof

Costs of Floors and Ceilings

Applying Skills

Find the area of each polygon.

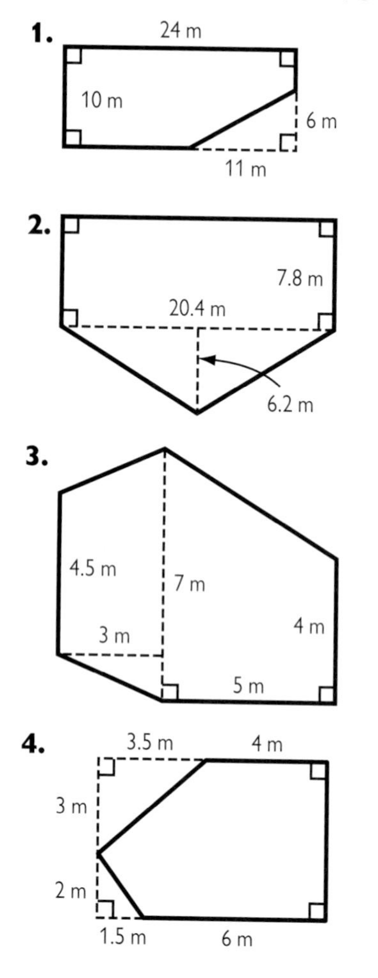

5. Find the cost of materials for the polygon in item **1** at $97 per square meter.

6. Find the cost of materials for the polygon in item **2** at $34 per square meter.

7. Find the cost of materials for the polygon in item **3** at $86 per square meter.

8. Find the cost of materials for the polygon in item **4** at $120 per square meter.

Extending Concepts

9. At a floor covering store, the prices for carpeting are given in square yards. The prices for vinyl floor tiles are given in square feet. Which would cost less: to cover the floor of your 10 ft by 12 ft bedroom in carpet costing $25 per square yard, or in tile costing $5 per square foot? How did you decide?

10. Suppose that your bathroom ceiling is covered with 9-inch-square tiles. You wish to remove the tiles and paint the ceiling instead.

    a. If you count 8 tiles on one side and 10 tiles going the other way, what is the area of the ceiling in square feet?

    b. Before painting the ceiling, you plan to install a new vent measuring 12 in. by 18 in. How much will it cost to paint the ceiling if paint costs $2 per square foot?

Making Connections

11. In a classic Japanese house, the floor is covered with tatami mats measuring 6 feet by 3 feet. The figure shows a floor plan for a Japanese house. Assuming that each rectangle represents a tatami mat, find the floor area.

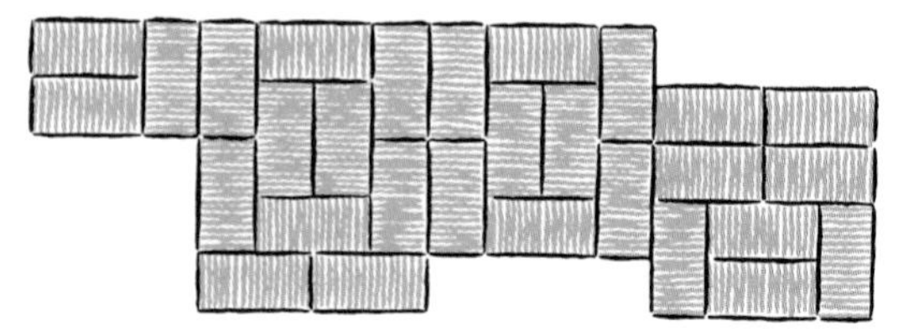

Homework 10

Adding the Cost of Labor

Applying Skills

If building materials cost $50,000, find the labor cost for each city.

1. Glenwood: 40% of materials cost

2. Duluth: 45% of materials cost

3. Bakersfield: 50% of materials cost

4. Sarasota: 65% of materials cost

Find the missing costs for each house if labor costs are 70% of the cost of materials.

	Cost of Materials	Cost of Labor	Total Cost to Build
5.	$39,560		
6.	$58,344		
7.	$71,875		

Check each calculation. First estimate whether the total cost is reasonable and tell how you decided. If the total cost is incorrect, give the correct answer.

	Cost of Materials	Cost of Labor	Total Cost to Build
8.	$43,000	50% of materials	$60,506
9.	$60,260	35% of materials	$81,351
10.	$50,390	60% of materials	$85,624

Extending Concepts

11. Suppose you are an estimator checking figures on a project. The cost of materials is $39,790. Labor costs are 53% of the materials cost. Estimate how much you would expect labor costs to be. Explain.

12. Better Builders Architects calculate the cost of labor by adding a fixed fee of $8,000 to 45% of the cost of materials.

a. If materials cost $40,000, how much would labor cost? How much would labor cost as a percentage of the cost of materials?

b. If the cost of materials is higher than $40,000, do you think that the cost of labor, as a percentage of the cost of materials, would be greater than, less than, or the same as your answer in part **a**?

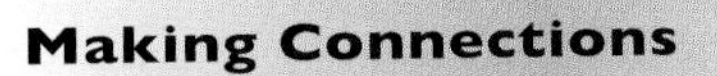

Making Connections

13. This graph compares the average price per square foot to rent an office in four different cities. What would it cost in each city to rent an office with 800 square feet for 1 month?

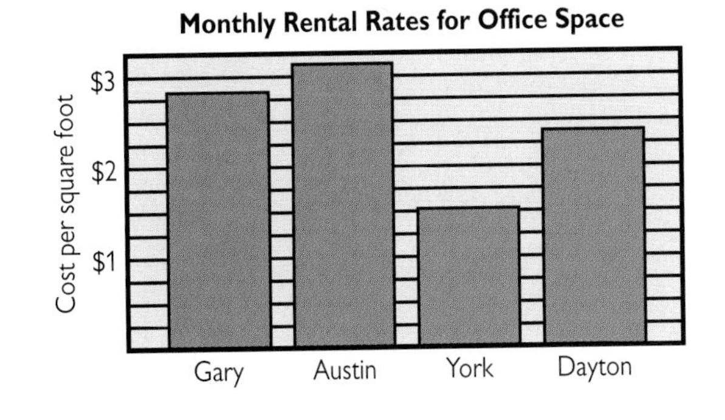

Homework 11 Designing Within a Budget

Applying Skills

Use the chart to find the total cost of materials for each house.

	Ground Floor	Roof	Outside Walls	Ceiling
1.	$9,920	$6,930	$8,750	$5,950
2.	$11,160	$8,227	$9,625	$7,350
3.	$15,500	$11,715	$10,780	$10,325

For the following houses, labor costs 55% of the materials cost. Tell whether each house is over or under the target total cost.

4. materials $49,204; target $75,000

5. materials $64,500; target $100,000

6. materials $101,320; target $155,000

7. materials $76,330; target $116,000

8. Suppose that you want to cut the cost of the house pictured in this floor plan. You do not want to change its total area or wall height. What can you do to keep the same floor area in your house at a lower total cost? Explain why your changes would reduce the total cost of the house.

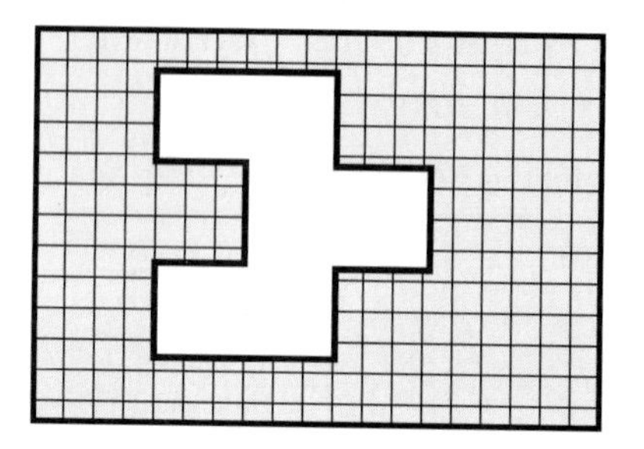

Extending Concepts

Refer to the rectangular floor plan for items 9–10.

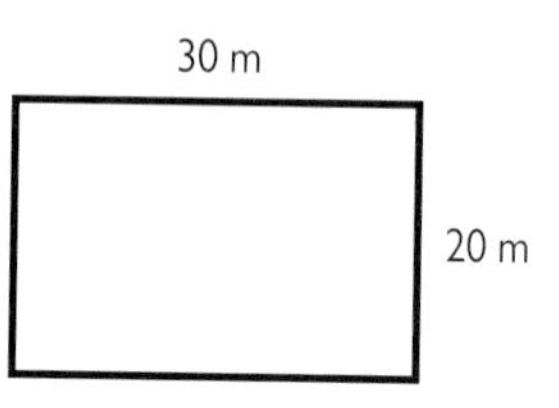

9. What happens to the area of the floor plan if you double the dimensions of all sides? Explain.

10. What happens to the area of the floor plan if you double only the length? If this is different than what happened in item **9,** explain why.

Making Connections

11. The lowest recorded temperature in Maybell, Colorado, is −61°F and in Tallahassee, Florida, it is −2°F.

a. In which city would you expect material costs to be higher?

b. If labor costs are about the same in both cities, in which city would you expect the cost of labor to be a lower percentage of the cost of materials?

12 Homework

Touring the Model Homes

Applying Skills

Find the cost per square meter of floor space to build each house. Round your answers to the nearest dollar.

		Total Area of Floor Space	Total Cost to Build
1.	Jose's house	84 m^2	$55,000
2.	Lee's house	175 m^2	$82,000
3.	Mei's house	620 m^2	$275,000
4.	Tom's house	105 m^2	$64,000

5. Considering total cost to build, order houses 1–4 from least expensive to most expensive.

6. Considering cost per square meter of floor space, order houses 1–4 from least expensive to most expensive.

7. This floor plan shows how the Lora family plans to add on to their house. The architect estimates that the addition will cost about $500 per square meter to build. What is the cost of the new addition?

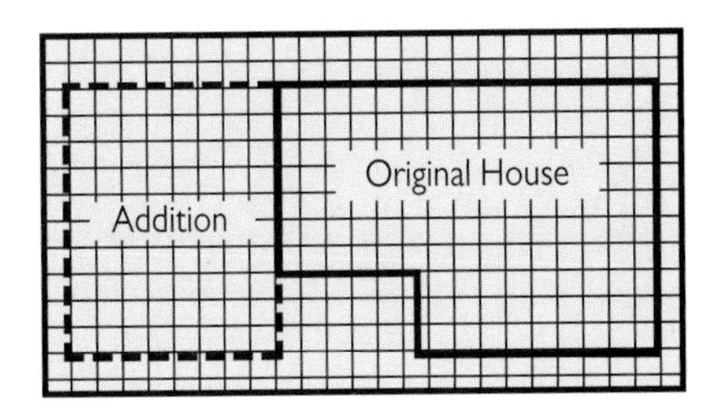

Each square represents one square meter.

Extending Concepts

8. Which surface would you choose for a driveway with an area of 300 ft^2?

Surface	Cost per ft^2	Lifetime
Brick	$7.50	20 years
Blacktop	$2.25	5 years
Concrete	$3.50	10 years

9. Read these ads. Which house gives you the most for your money? Tell why.

Move right in!
Just $165,000 buys a 3-bedroom, 2-bath house on a large lot close to good schools and shopping. Large garden and garage. 1,760 square feet.

Clean and neat!
Nice home in a tree-shaded neighborhood. 1,042 square feet; 2 bedrooms, 1 bath; new kitchen appliances and cabinets; carport. $95,960

Writing

10. Answer the Dr. Math letter.

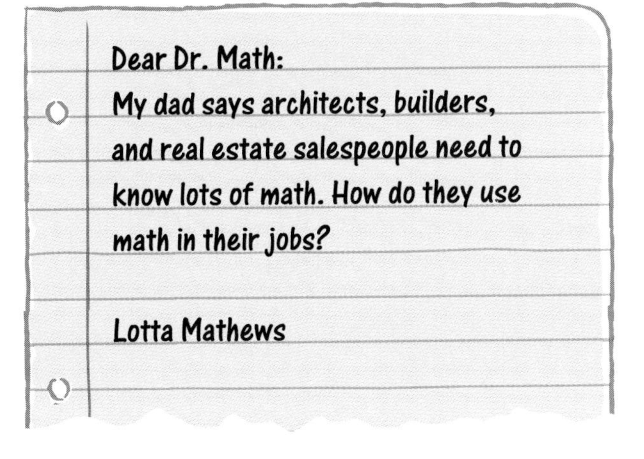

Dear Dr. Math:
My dad says architects, builders, and real estate salespeople need to know lots of math. How do they use math in their jobs?

Lotta Mathews

The McGraw-Hill Companies

This unit of MathScape: Seeing and Thinking Mathematically was developed by the Seeing and Thinking Mathematically project (STM), based at Education Development Center, Inc. (EDC), a non-profit educational research and development organization in Newton, MA. The STM project was supported, in part, by the National Science Foundation Grant No. 9054677. Opinions expressed are those of the authors and not necessarily those of the Foundation.

CREDITS: Unless otherwise indicated below, all photography by Chris Conroy.

141 ©1994 TSM/Peter Beck.

Send all inquiries to:
Glencoe/McGraw-Hill
8787 Orion Place
Columbus, OH 43240-4027

ISBN: 0-07-866812-3

5 6 7 8 9 10 058 09